AF837047

Greetings to you
and
Best Wishes
Harold E. Wagner

BARNS ON PARADE
Across America

by

HAROLD E. WAGNER

Photographs and Illustrations
by Harold E. Wagner
(except as individually noted)

BARNS ON PARADE
Across America
by Harold E. Wagner

 Distributed by H & O Pioneer Barns, 508 E. 3rd, Grandview, WA 98930 and Northwest Publishing, 1130 Columbia Ave., Sunnyside, WA 98944.

Prepress arrangement by Northwest Publishing, Sunnyside, Washington.

Printed in the United States by Adams Press, Chicago, Illinois.

First Printing: July 1993

Library of Congress Catalog Card Number 93-71757

ISBN
0-9637155-0-X

Published by H & O Pioneer Barns, 508 E. 3rd Street, Grandview, Washington 98930. Phone: (509) 882-1799

Note: In your reading this book on barns, if you discover any discrepancies, please notify H&O Pioneer Barns at above address or telephone number.

To mail-order additional copies of this book, please see the order form at the end of this book.

"READ ...PONDER...REFLECT..

ENJOY WITHOUT NEGLECT"

My sincere appreciation and love to my beloved wife, OLIVE WAGNER

...Who has been my partner 46 years

...Who believed in this barn project 10 years

...Who has travelled with me thousands of miles

...Who patiently stood by during lengthy interviews

...Who kept a running account of daily events

...Who taught English and edited this writing

The Cover Photo

ONCE A BARN

NOW

THE LEWIS ALUMNI CENTER
Washington State University
Pullman, Washington

The Great Transformation

~ The Legacy ~

Stand By - Read all about it!

Chapter Four, Pages 10-35

Cover photo courtesy of
The Lewis Alumni Center,
Washington State University

Dedication

I, Harold Wagner, the author,

herewith dedicate

this book

to anyone interested in barns:

Farmers - Architects - Engineers - Builders

Galleries - Libraries - Book Stores - Coffee Shops

Artists - Photographers - Writers - Publishers

Future Barnologists

AND

READERS

AUTHOR'S PURPOSE:

1. To raise awareness of the many barns being destroyed each year.
2. To encourage maintenance and/or rehabilitation of the barn for other uses, as others across America have had the foresight and courage to undertake this kind of preservation.
3. To record historical data and many interesting stories about the lifestyles of the pioneers who built and utilized these earlier landmark structures.

These pioneer barns have unique designs and character for all to see, who pass by with curiosity and admiration.

TABLE OF CONTENTS

FOREWORD

by

Friends of the Author

JEFF ANDREWS

MAURY LANGE

CLIFF and MICHELLE LEWIS

HAROLD E. WAGNER, AUTHOR
Grandview, Washington

The following is an excerpt from an article published in the Tri-City Herald, Kennewick, WA, on August 2, 1992. The story later was carried by Associated Press by various newspapers throughout the United States.

By Jeff Andrews

GRANDVIEW, YAKIMA COUNTY, WA — In 1983, Harold Wagner went back to Osborne County in Kansas to see an old barn he helped build 50 years earlier.

It had been torn down.

"That seems to be the American attitude, the 75 year-old Wagner lamented. If it's old, get rid of it."

That experience ignited a passion within him to change attitudes about old barns. It was the start of a save-the-barn campaign that has taken Wagner, a retired Grandview real estate broker, down thousands of miles of country roads in the United States and Canada.

Among his many trips, he traveled through 30 states during a five-week journey in 1990. In May 1992, Wagner went to Colby, Kan., to watch a 66-foot wide barn moved 16 miles to a museum there.

"He spent hours out there," says Inez Mountford, a researcher for the Prairie Museum of Art and History in Colby.

"He took notes, he took pictures and he visited with hundreds of people."

Wagner says he's photographed between 1,300 and 1,400 barns since 1983, and he has a story to tell about most of them. Many are included in this book.

Wagner's neat two-story house in the Lower Yakima Valley is filled with watercolor and oil paintings, many of barns. He painted many himself. Rows of barn pictures are lined up on a table and two filing cabinets are filled with notes, literature and pictures of barns.

His research is methodical, even a science, the self-described "barnologist" says.

Wagner has identified 21 different types of barns based mostly on roof designs. Barns come in four different shapes: rectangular, square, round or L-shaped.

Round barns were built primarily to allow central feeding of animals: more animals could be fed using less space. Wagner recalls an anecdote about a farmer who said he built his barn round to keep the hired hands from urinating in the corners.

When on the prowl, Wagner also examines the condition of barns, their superstructure, the types of materials used and the construction techniques.

Many East Coast barns were made of stone or brick, while West Coast barns mostly are wood, he said.

"The insides reveal a tremendous amount of artwork, the artwork of construction." he said.

Wagner has built a formidable network of informants over the years. Each week, he gets letters from throughout the country with information about where an interesting barn may be found, or containing the history of a particular barn.

When he sets out on a trip he determines a number of barns he's going to visit, but along the way looks for others.

Wagner is often accompanied by his wife, Olive, to whom he has been married since 1947. She often takes notes as they drive along back roads.

But barns are getting harder to find.

"They aren't used much anymore," researcher Mountford says. "They're either falling apart or getting torn down. It's nice someone is gathering information on barns, so there will at least be records so young people will know about them."

* * *

At very special times in life generations we are blessed with the opportunity to reach across the chasms of different eras, needs, and realities to touch each other with the flames of each others' dreams. Harold Wagner has done that for me...not just sharing his ample vision (faith, family, politics, land and even BARNS!) with me, but also allowing me to share mine with him. For this I am forever in his debt.

BARNS ON PARADE ACROSS AMERICA is a book about a significant American icon; it is also a photo-essay of one man's odyssey in making a "whole fabric" of his life...Christian pastor, husband/father/grandfather, farmer, REALTOR, craftsman, civic volunteer, spokesperson, and author. Harold's cherished barns are silent sentinels tying the past, present, and future of our land with the values that have survived, those that are changing, and those that have passed into history.

Some of their stories have been lost forever; perhaps some should never have been built. but, our barns are the link not only with our past but also between our land, our productivity, and our stewardship.

Harold has been an avid champion of private property rights. His barns reflect the very souls of those who dared to challenge the land to give its very best over the years. **Read this book and ask yourself if you have cherished the land sufficiently to defend it.**

MY FRIEND, HAROLD WAGNER, FAITHFULLY THANKS GOD FOR THE GIFT OF THE LAND AND CHALLENGES ALL OF US NOT TO BE SO HASTY IN DISREGARDING THE PAST. THANK HIM, YOURSELF; **JUST READ THIS BOOK AND TOUCH SOMEONE ELSE'S FLAME!**

Thanks, Harold,

Maurice A. "Maury" Lange
Friend, student, REALTOR
March 18, 1993

* * *

The author of this book, Harold Wagner, is a well known friend here in Grandview, Washington. He has been a customer of Grandview Lumber Company for many years, carefully selecting lumber, hardware and tools for his projects. Mr. Wagner seems to have an excellent knowledge of the different kinds of wood/lumber and preservatives. His workable knowledge of construction requires little consultation for either small or large projects.

Cliff Lewis
Michelle Lewis
GRANDVIEW LUMBER COMPANY
March 13, 1993

Chapter One

INVITATION TO A PARADE OF BARNS

This is an invitation to a parade of barns. First, let us check in on a few other parades. Who among us, whatever our age, wouldn't drop everything to watch a parade. It's our up bringing, our desire to see the unusual displays, and to see funny things that occur, during a parade.

#1

Golden days of the slower pace.

#2

Photos #1 through #5 depict a parade in the past in Osborne, Kansas.

#3

#4

#5

Not all interesting parades are found in the towns and cities, but rural areas of America offer some fascinating scenes of farm equipment parading through the fields.

#6

Heading Kansas wheat. This author was the driver of the team pulling the header-barge

#7

Wheat combine, pulled by a tractor (older style) in the 1920's and 30's. This author was the tractor driver. The combine cuts off the wheat heads and flails out the grain. Faster and more efficient than hand beating.

Community celebrations in rural America often times parade equipment through fairgrounds or parks, such as this Berthusen park (#7), two miles or so northwest of Lynden, Washington, during the Puget Sound Antique Tractor and Machinery Show each August.

Parades, even to this day, have never eradicated the horses. If anything, horses prominently have their place in nearly every parade, and rightly so, because of their natural live power, their variety, their attractiveness, and their driver or rider's pride.

#8

As it was years ago. Horse power source for transportation; slow, but it sure beat walking.

#9

Early 20th century changes to kerosene and gas engine power, gradually for two decades or so.

This was "the modern" truck at one time to haul livestock and grain. Today, it is a valuable antique and show vehicle. It is more valuable than the average new car today.

#10

#11 Not much power, but it sure gets attention!

#12

These big steamers were a boy's delight to ride. It made you feel like a man.

#13

Tractor advancement.
Four wheel drive power for those big jobs.

Chug-a-lug. The parade is on four
tractors, small to the large.

#14

Parades come in many places, styles, and varieties. Thinking about barns, can they really be considered to be on parade? Well, when you finish this book, you will feel that you have seen a parade of a different kind. At least, I hope that you might think of it in this manner. It will have more impact on your interest and memory. In fact, I venture a guess - you will from this day look for barns as you travel, have a greater appreciation, and just maybe you will take your camera along and photograph some. I did! Others tell me that they did.

You will find a variety of styles, shapes, sizes, conditions and uses (if any). Look at the location or the setting; look at the architecture; look at the engineering for the strength, stability, and utility; imagine yourself as the builder, climbing high on the framing with nails in hand and driving them with the hammer. In 1933, this author, at high school age, helped in the construction of a new gambrel roofed barn in our Kansas farm. The height - was it scary? You bet! Determination, practice, confidence and a relaxed body helps. Years of experience are beneficial, until you become an old man like me with weakened muscles and unsure legs – you just don't climb those ladders anymore. However, over the years, this author had known, visited, and admired many builders. He has commended them.

While we are on this subject, if you ever get to visit the Empire State Building in New York City, or any other tall structure, as I have, when I lived in New York City, you cannot help but appreciate those builders and those superstructures. Native New Yorkers always can recognize newcomers. They tend to look up and bump into people. "I'm sorry" becomes a part of your vocabulary for awhile, that is, if you respect others.

Respecting others – here is a legendary story told in New York City, years ago. Picture this scene: Two young ladies were walking down 42nd St. near Broadway. One of them was loosing her panties, due to weak elastic. She bent over, quickly stepped out and placed them in her purse. Three sailors coming from the other direction, quickly looked up and pointed to the top of the Times Square building. That diverted all eyes off the ladies. The ladies, in all politeness, said, "Thank you!" Now, that gesture was respect in action, wasn't it?

Getting back to the main subject, why should anyone but a builder or a farmer be interested in any kind of a barn where there is manure, debris, dust, odd odors, etc.? That question has been asked of me many times, and you may feel like asking it.

Check it out! These pioneer barns become land marks on the rural landscape - some are even still located in towns and cities, but each year, many are destroyed. Barns originally were built for hay and grain storage, for the housing of livestock and for some farm equipment.

In addition, many youngsters found the barn a great place to play hide-and-seek, as well as other games. Many a haymow had the hay pushed to one side, the floor cleaned, friends invited in for a "skip-a-dee-doo," the big barn dance with a "caller" and music; Chicken-in-the-Straw, Two-Step Promenade, Four Squares Around and many other movements. After all, the barn was usually the largest facility in the community to hold parties and dances in the by-gone days. Some barns still have band platforms, and some still are entertainment centers.

It's not uncommon to hear amusing stories about barn activities. One such story was written to me in a letter, 1992, by Liz Johnston of Kennewick, WA. Her family visited a huge barn at Sequim, WA, where the father of a young daughter had put up a rope swing in the top of the barn. Liz says "that (it) took about 45 seconds to go from one side to the other." Then she adds, "Our children and I don't pass an old barn without a smile and the question 'Is there a swing in the loft?'"

Many people across America send me informative letters for which I am grateful. Another letter writer, Pat Shuyler of Lebanon, OR, wrote me about an old barn, where someone (apparently with a Geiger counter) came into the barn and "in the doorway of this old barn the 'needle' went berserk. They dug up a jar of old $20 gold pieces." Does that excite you to do likewise? Maybe I should invest in a Geiger counter. It has crossed my mind, but then that would consume too much of my time when I'm photographing barns. I must add, I love to get these letters from people with interest in barns, other homestead structures, such as windmills, silos, cribs, and sod houses. A return thank you letter to you is my practice. What story and/or photo do you have to voluntarily share?

Before we move on to the next chapter, let me leave you with this thought about barns. A barn is a place where you learn so much knowledge about birth, life and death. Events in and around a barn are so varied and extensive that one can never begin to imagine, until you have been a part of the daily goings-on.

Chapter Two

THE BARN THAT WASN'T THERE

Previously, you'll recall a mentioning of my involvement in helping with the building a gambrel roofed barn in 1933 when a youth on the family farm in Kansas. My older brother, Orval Wagner, also worked on the barn. Here is a photo of this barn, which was taken shortly after completion of the construction.

#15

Photo courtesy of Orval Wagner

This gambrel roofed barn replaced a gable roofed barn which was ready for the bone pile – it was razed.

This new gambrel roofed barn contained several horse stalls, a birthing pen, two grain bins, a wide stairway and a spacious haymow. Why there was no cupola built for good ventilation for new stored hay, I don't know. Cupolas are often built and with variety of interesting styles. Most cupolas attracted sparrows, pigeons and a few other birds. Speaking about birds, swallows liked the barn eves for nesting. Owls liked the haymow area, usually a darkened corner for nesting and to spot mice for their diet. Obviously, we would never kill a barn owl, because of their benefit and their peculiar characteristic and movement. A very interesting bird species they are.

In 1983, after many visits back to the farm during the intervening years, that barn wasn't there anymore. What a shock! It was a land mark, which could be seen from various location. That made me wish I had taken more pictures, and that started me on a barn research and photographing of barns across America, covering enough miles that would have reached around the world. Chasing after barns – I'll bet there is someone out there who thinks I'm crazy. This reminds me of the "legend" story about a round barn, where a newly hired man went crazy, running around inside to find a corner to urinate. Yes, that would be a problem, don't you think?

When I found that big red, gambrel roofed barn had disappeared, that was not all there was to my emotional trauma. Our home, concrete cave (storm cellar), outhouse (the "privie", so-called by farm families), smoke house to cure meat, granary, and cow shed were all gone. Can you feel how I felt? The next photo (#16) is of our former home and the privie in the background to the left side of the photo. That was the "recycling house" (you might call it) for the Sears Catalog. The recycling was necessary, even back in those days, and it did work very well for those catalogs that were outdated. Catalogs – the women and children liked them for use. As for the teenaged boys and men, they preferred the fresh corn cobs from the crib! - No joke. There is an old legend out there about those corn cobs: It generally required three cobs – two red ones and one white one. How come - why? Well, a red cob was used first, then the white cob was used to see if you needed the other red cob. We always raised lots of corn, so cobs were plentiful, since we shelled our own corn and saved the cobs. Now, it's time to show you the house and the privie photo.

#16

Wagner family's 3-bedroom home near Covert, Kansas. Also see the cistern, cave (cellar) and outhouse.

If you would like to do some research on my background, I would advise you not to write to the Chamber of Commerce or the post-office in Covert, KS. The two stores, post-office, bank, garage, cream station, central telephone, two churches, and barbershop are gone (long time now). That big stone marker on the previous bank corner cannot help you at all.

The "great depression" in the 1930's; plus the long drought; dust storms and invasion of grasshoppers, caused a big panic which devastated the farms and ranches. So many farmers lost their places, took out bankruptcy, held auctions, and then moved on west to California, Oregon and Washington. My parents desperately hung on until 1942, when they finally gave up to come to Washington, the Yakima Valley that "flowed with milk and honey" (the biblical concept of prosperity).

The farmers who stayed, sacrificed and struggled to make a go of the farm were the younger ones with a lot of vigor and determination to keep their dream alive. This author's dream, at that teenage period of his life, was to get a college education and a career – not farming at that time.

Take a good look at this next photo. See if you can see any buildings. Well, just maybe you see something in a heap to the left side of the tree located to the left side of the photo. This pile or heap with a growth of weeds on it was the cave (cellar). It was and still is the only remaining identified object, where the home and barn represented a thriving homestead.

It was nice of someone to leave the cave (cellar) to identify the homestead site where a family labored hard for a living for several decades. #17

Photo courtesy of Orval Wagner

If only that cave (cellar) could talk, I'm sure it would reveal many more stories than this writer has ever heard. Permit me to share a couple stories. When those terrifying storm clouds gathered with the possible appearance of a tornado, you dropped everything and rushed to this cave for protection. One time a storm took off the porch, but left the house standing. As for that cave, it was safe and you could exist there for days, if necessary. It was well stocked with potatoes, apples, canned vegetables and meat; plus it had a vent pipe.

What disappearance took place with our homestead buildings, the duplications occurred in the mid-western and western states of America - sad but true. This disappearance of the homestead has continued progressively to this day, due to the farming economy and young farmers heading to the city for good paying jobs.

This next photo illustrates very vividly the movement from the farm. In a very short time, these buildings also will be gone, followed by the closing and boarding up the stores, school and churches in the nearby town. Is this the kind of legacy we want for our nation? How can we feed ourselves and the world?

#18

My photo in 1992 of another Kansas homestead on its "death bed".

So that you don't think that farm migration is only taking place in middle America, let me show you another photo (#19) taken near the town of Fox, OR.

#19

Where have all the people gone?

WHY?

My research is very extensive in the United States and parts of Canada, showing the death of farms, ranches, towns and villages. The main contributing factors are weather conditions, growing bureaucratic regulations, and the outdated system of agricultural marketing economy. This outdated marketing economy is where the farmer goes to the market with his products/commodities and asks, "What will you give me?" He lacks controlling power to get a price for his product that will cover his costs of production, his risks of weather conditions, and the restrictions of the strangling bureaucratic regulations. Yes, and what about having a little of something left to feed and clothe his family? Think about it – Would you like this situation? Why do you think farm youth are heading for the city, wanting your job? But the most serious problem, as this situation increases, is the threat to our food supply.

Are there any answers, any solutions, any hope? If you want food on your table, if you want farmers to stay on the land, if you want to save your jobs, plus many other if's we could mention at this point, then we must find answers and better solutions. Only then can hope be restored and sustained.

What about the weather? It is out of our control, isn't it? What about regulations? They can be moderated and minimized by better public understanding and cooperation of those directly involved. What about the system of the marketing economy? "Agricultural bargaining power must be universally allowed and applied" in all areas of farm product marketing. General farm organizations have come into existence to serve certain needs of the membership. Coops, associations, and a few other bargaining groups have had limited successes. But there is still the need for "universal bargaining power" in the agricultural industry.

Farm subsidies by the Federal Government have only involved a limited number of farm commodities (eleven, if I remember correctly). Increasingly, the government is phasing these subsidies out of existence. This being the case, it is all the more reason for a "big change" in the marketing system for agricultural commodities in the form of universal bargaining power. This is the "year of change," this year of 1993 and following. Our politicians have told us this - that changes must be made. Who among us would disagree?

If agricultural marketing maintains the status quo and not make the necessary changes, we will experience even greater devastation down on the farm.

Agriculture has experienced great changes in the area of mechanization and production practices, but is it enough? The answer is NO! The agricultural marketing system must be changed.

Chapter Three

BARNS IN WAITING
(No Job)

"Time has come for the birds to sing."
Solomon 2:12

Time has come for us to find a good place to sit and watch the parade of barns as they come along our way.

In this writer's early years on the farm, looking up at the full moon one night, my dad asked, "Do you see that man in the moon?" My answer was, "I think so." Years later, I learned that the Chinese thought that object was a rabbit (or maybe that was a robot) – whatever. Objects often appear differently to different people. Different experiences, knowledge and interest have a bearing on our perception and interpretation of the objects we see. We find this true, as we look at barns.

To me, the pioneer barns, as constructed with the most simple hand tools, are the "masterpieces of yesterday" whatever their style, shape or condition. During the past nine years of my research, invariably I was asked this question. "Why are these barns being neglected, falling or being destroyed? Assuming you read Chapter Two, you begin to get some insight on the problem. Years ago, buffalo were once plentiful, unprotected, and, except for a few here and there, have disappeared from the scene. This scenario you'll find illustrated as this parade of barns appears before you.

And here they come. This first one, we might call "ol' Smiley."

#20

In Idaho it has served some good years. It is not ready to give up yet.

Photo courtesy of DEAN SQUIBB

Barns are like humans. They need a lot of regular grooming and attention to withstand the storms of life. But sometimes a storm will devastate beyond replacement.

#21

Iowa raised – Great place! So weak – Can't go any further!

#22

A resident of Nebraska that couldn't lay down. Restricted by a corn chopper in the left corner. (You can see the bulge.)

This huge granite stone barn (#23), known also as the "North Family Barn," is about 300 feet long and five stories high. Who knows of a larger barn? Please let me know!

The BIG One

#23

Mount Lebanon Shaker Village Barn, New London, NY.

This barn was gutted by a fire in September of 1972, as were other barns in the northeastern part of the United States. The west long wall had to be stabilized by five steel heavy beams. Since the fire, trees have grown up within the barn walls. When my wife and I visited the village and photographed the barn in September, 1990, those trees had grown past the third floor markings on the walls. The setting and the barn, as it is, causes you to stand in awe - it is so huge. At one time the barn was considered the largest stone barn in North America. To restore the barn's floors, roof, windows, clean-up and stabilization, I have heard of costs ranging as high as two-million dollars. Who has that kind of money for such a restoration and economic need?

The village has many other buildings (25 total in all), some occupied by the Darrow School in New Lebanon, New York. Restoring and preserving these Shaker buildings has become a project of the University of the State of New York.

The Shaker settlement was founded in the 1780's by a young religious group under the leadership of Ann Lee, an English woman who came to America in 1774. They believed in confessions, celibacy, separation and communal living. Their brochure given to me entitled "Mount Lebanon Shaker Village" indicates that by the mid-nineteenth century the Shakers had eighteen societies in the Eastern United States from Maine to Kentucky. The Shakers were very industrious people and have left a legacy of making some of the finest furniture produced with simplicity, utility and uniformity.

Knowing Tim Rieman, chair and furniture maker, whose studio is in the North Family Granary building (across the road from the barn), and knowing some of his carefully crafted furniture, I applaud him for his fine craftmanship. Collectors are buying this furniture as prized keepsakes.

As for this Shaker barn being restored, it would be a very worthwhile project, saving it for use again. But twenty-one years, thus far, and economics being what it is today, who will step forward and be that generous philanthropist?

Today this huge classic barn, even left as a shell of stone, draws much public attention. Therefore it deserves much commentary to keep the historical data and value alive. It stands as a monument to the great achievement of the Shaker family.

An idle Virginia barn just hanging around with the trees and bushes. No Job!

#24

#25

This barn in Washington is trying to hang on to serve one horse and maybe some sheep. Not far away, Mt. Adams is standing watch.

This barn enjoys good air circulation (Washington).

#26

Kansas round barns have nothing to do except to die. #27

#28

RIGHT: A Washington corral holds nothing but weeds – a common sight on many farms tody.

BELOW: This Missouri barn doesn't like storms. They hurt!

#29

#30

New Jersey stone barn awaiting the next swing of the big ball.

#31

Idaho barn with an aching, weak back. Beyond help of a good doctor.

Photo courtesy of
Dean Squibb

#32

A Washington barn, gable roofed with two lean-tos. Main structure siding is board and bat.

#33

Old drayage barn with board and bat siding with a pioneer village frontage.

Photo courtesy of
Fern Eckroat

#34

California barn still in service, plus an unusual square silo in front.

Photo courtesy of
Fern Eckroat

#35

Washington barn waiting to be razed, relocated, and reconstructed for a farmers' market ten miles south. A great tomorrow for this barn

Ila Timson, a former resident of Nebraska now living in Kansas, wrote an informative letter about this barn (#36). She reported that this red tiled round barn had another

#36

A Nebraska round barn, except for the roof, withstood a terrible strong storm. A good reason for building round barns.

barn like it nearby, also two silos built byBob Radford and sons in 1921-23. The barns had hardwood haymow floors, used for dancing and basketball games by the locals (community). In 1947, the Walter Younkin family bought the farm. In 1952, a hard wind toppled one silo on top of one barn, destroying both barn and silo. In 1984, the top of the remaining barn and silo roofs were blown off.

In 1992, this writer took the above picture and met the grandson, Gary Younkin, who operates the farm and he told me much of the same story. He reported the barn size at 73 or 75 feet diameter and 53 feet high with center wells in both barns for easy and ready watering of the animals.

#37

North Dakota barn, like a tired sway-backed horse, before going down. Other states, including Washington, have these "downers."

These barns attract a lot of attention, especially from photographers and artists. Why? It is a "last minute chance" to preserve not the barn, but to preserve the "unique image," as if it were a fallen soldier. Eloise Ogden, staff writer for the Daily News of Minot, North Dakota, calls these barns "silent sentinels on the lone prairie." What a nice creative line - I like it! In her writing, she raises some very serious concerns, concerns that I share - one of the three purposes for this book.

In the Sunday, January 26, 1992, issue of the Minot Daily News in the Prairie Lifestyles section, Ms. Ogden wrote a very lengthy, informative article on barns. Here is what she had to say. "Those big barns on North Dakota farms and ranches are gradually vanishing from the countryside. And with their disappearance comes a loss of North Dakota history. Once the center of activity for the farmers and ranchers' work, barns were also a social place for dances." How true! That statement could be applied to states across this nation. It is a symptom of our economic times, as I have tried to illustrate in this chapter.

When you read Chapter Four, you will find many illustrations of active barns, barns still serving shelter for animals, hay, grain, and equipment. You'll find many barns across America converted and rehabilitated for other purposes. In closing this chapter, I prefer to turn to poetry. I enjoy poetry and prose, and I hope you do likewise.

PIONEER FARMER AND HIS BARN

This rolling land is a weedy old farm,
And there stands an old decaying barn.
The farmer is very poor with little money,
And it makes him feel sick and crummy.

Running out of hay he lost his milk cow,
And he is in a terrible bind right now.
He raised a big family and they are gone.
His wife left with a bum, that seemed wrong.

Life has been very long and not very fair,
But a friend brings him food in loving care.
Being old and lonely he often feels so hurt,
Expecting he and his barn will turn to dirt.

Harold E. Wagner
Writer/Poet

Some people like poetry, while a few others prefer straight talk or something else. Most poetry can reveal truths about a situation. Personally, this writer enjoys many kinds of poetry and prose.

> The Bible speaks of a "time to be born
>   and a time to die."
> Between these two times is a time
>   to live, be happy and sometimes cry.

Chapter Four

BARNS ALIVE – ON DUTY

This chapter showcases a much larger selection of barns. These are the active ones. Some are still housing livestock, as most barns originally were built for this purpose. You will find, today, many barns rehabilitated and converted to other uses, as you see these barns parade before you.

It was Thomas Edison, who said, "An idea is something that won't work unless you do." In my lengthy research, I have been very amazed at the extent of ideas proposed and the work involved to save these pioneer barns.

Now, let's sit back and enjoy the next division of the parade: the "active" barns.

#38 Washington State University (W.S.U.), Pullman, Washington. Here stands a beautiful, exquisite and elegant gambrel roofed structure, known as the Lewis Alumni Center (West entrance [front view], facing the main part of the campus.) This building is 140 feet long.

#39 This east entrance faces out to the parking area. It is attractively landscaped. The photo gives you a view of the two extended wings with lower roof lines.

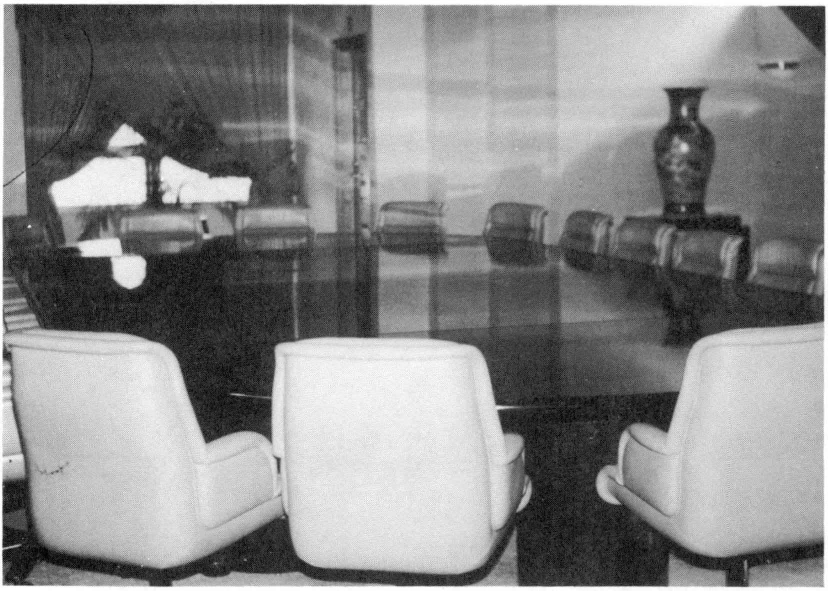

#40 ABOVE: W.S.U. Regents Room located in the Lewis Alumni Center. It is attractive, comfortable and serviceable. **#41 BELOW:** The W.S.U. Lewis Alumni Center gives us an overview of the superstructure. Regents Room is in the wing closest to us in this photo. The total structure flooring represents approximately twenty-five thousand square feet.

#42

Take note of the ceiling structure. It was maintained to give evidence of the barn, known as the animal science beef barn. This lounge area was once a haymow.

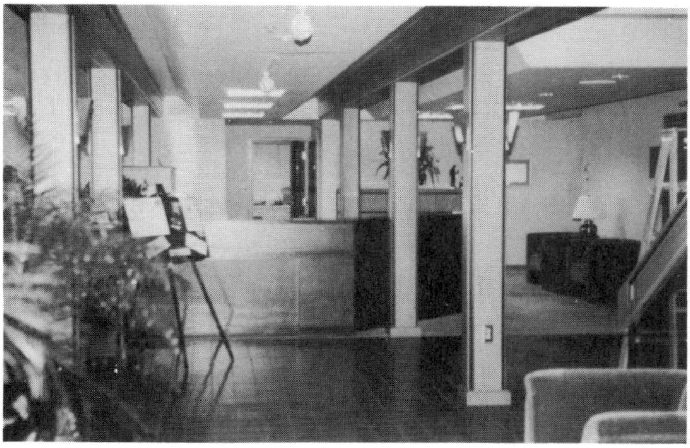

#43

Alumni office work area (first floor), private plus adjacent offices and meeting rooms.

Before we approach the Cougars, I should tell you that those two silos shown in front of the Alumni Center are not there to be filled with corn. Someone might see it as a "corny" idea, but let me tell you, when you need "to go", there is nothing more practical. Each of those silos are practical. Left one is a two level women's restroom and the men's is on the right of the entrance. The fixtures are curved to fit the curved walls. Very nice!

#44

W.S.U.'s bronze Cougar sentries are positioned at entrances to welcome visitors. The Cougar is W.S.U.'s mascot.

Why did I choose for the cover this Lewis Alumni Center on the campus of Washington State University, Pullman, WA? Several reasons convinced me to make this choice (I had seriously reviewed other photos with stories for the book cover). But this structure was originally a barn that was used by this Land Grant College for science purposes, which made a tremendous advancement in animal husbandry, helping farmers to be better farmers and ranchers to survive on the farm and therefore result into a better food supply for consumers. For this one reason alone, I should give more attention in this writing. Second reason for my choice is that my wife, Olive, is a graduate of Washington State University (1941), giving her an excellent education for a career in the teaching of our youth and our own three children. She has stood by me in this research project on barns across America, assisting with the research in our many travels. Having been an English teacher, her knowledge has been most beneficial.

Third reason is the large size, gambrel style, excellent location and longevity of this structure. Fourth reason is the fact that when it was destined for removal in 1982 due to its dilapidation, enough individuals were visionary and looking for a spot to best showcase the University. Thus, the barn on campus was saved, reserving some of its structural image.

To accomplish this huge transformation from a beef barn to the Lewis Alumni Center by the administration and structural committee work, extensive plans were created for this unique University project. Promotion of the project was nation-wide among thousands of WSU graduates to raise the necessary funds. In all my years of researching barn transformation projects I have not found none that tops the approximately four million dollars ($4,000,000) of funds raised. It is a very remarkable accomplishment and construction got underway with completion in 1989 - just in time for the centennial. A great legacy has been created here. It really warrants a lot more space than I am allocating for each of these barns in this book. This Alumni Center becomes a monument to those accomplishments and contributions by those who were involved. WSU graduates can be proud of their association with their alma mater; its service to them, and their use of the facility when in Pullman.

As a barnologist, I am proud of WSU and the alumni for saving the campus barn and creating such a beautiful, serviceable and memorable Alumni Center. As a token of my appreciation, I purchased an engraved tile for the Great Hall reception area, as a gift to my wife, Olive Burnette Wagner, 1941. It exemplifies her pride in WSU and the Lewis Alumni Center.

I am pleased and thankful to Keith Lincoln, Gage Lyons and Ernest Weiss on the WSU campus for all their help.

On with more barns.

#45

Washington dairy barn with "monitor" style roof.

Photo courtesy of Allen Noble

#46

Grand Canyon pack-mule barn.

Photo courtesy of Fern Eckroat

#47

The llama barn. Horse Heaven Hills in Eastern Washington.

#48

Carnation Company dairy barn complex in the Snoqualmie Valley east of Seattle, Washington.

This Carnation Company (Photo #48) began in 1899 by E. A. Stuart, as a quality milk production and processing evaporated milk plant. Though the operation began in a re-arranged, abandoned hotel at Kent, Washington, later, in 1910 the operation was moved to a farm in the Snoqualmie Valley where it has continuously operated with on-going development.

When I first noticed this Carnation dairy, I was living in Seattle after World War II. Occasionally, I would take a tour out through the Snoqualmie Valley, which was a sentimental journey, especially after I had been out to the China Sea, for I loved country scenes with livestock - memories of the days of my youth on a farm. That is when we milked cows by hand, called "teating." Passing the Carnation farm, I was impressed with those black and white holstein cows. They were larger and seemed more docile and contented compared to some Jerseys I had milked, which were so temperamental and scared easily.

If you studied that picture of the Carnation dairy, you noticed a complex of barns, some alike with gambrel roofs and others with more modern style with lower roofs. In 1991, I had the privilege of visiting the Carnation dairy; permitted to walk through the barns, and was presented with two very interesting Carnation books, giving me a good history of the Carnation company and its progress through the years in the United States, as well as, in a number of foreign countries. It shows what vision, quality and leadership can accomplish.

Looking back at the last photo, you can see that the Carnation Company has preserved its earlier gambrel roofed barns, even in the midst of progress with more modern barn styles as we see in this next photo (#49).

Carnation's more modern dairy barn.

#49

With all of Carnations' barns grouped together, they form an unusual attraction for people interested in dairy history, taking a tour through the barns and looking at the cows. When I visited the Carnation dairy, I saw many families with their children coming to visit the dairy complex.

Here the children learn that milk is produced by cows - not by a grocery store. Why don't more families take the time to get an education of this kind? Why not learn values of family togetherness in exploring the world of activities around us? Why not leave behind those television shows which are life's negatives? Why allow such showings taking up our time to drag down our society? When our President calls for change, change and change, isn't it time to consider positive, up-change and change, isn't it time to consider positive, uplifting, and better educational methods, as we find here at Carnation.

Thanks to Carnation and companies like them, who provide information and curio shops with guides to point out the characteristics of the animals, the need and structure of the barns, etc. Any activity of this nature increases knowledge and fascination with life.

#50

Families coming to tour Carnation dairy.

Here are other active farms where the barns are used for animal shelters and work places yet today.

The owner of the barn pictured below, Marvin Tjoelker of Lynden, WA, wrote to me that the Monitor barn has five 12'x12' boxstalls and a tack room for bridles, saddles, etc. Monitor style barns are typical horse barns we find in many areas of America today. If you see this style of barn as you

#51

Well kept farm buildings in Washington. Left to right: Gambrel roofed barn, monitor barn and gable roofed barn. Good place for horses and cattle. Nice, attractive farm.

travel along, you can very well guess there are horses on the farm/ranch. In earlier days, they were housed in either the gambrel or the gable roofed barns. Dairy cattle were usually milked in a sanitized gable roofed barn. The gambrel roofed barn in the photo had been a milking barn, but now is reported to be a heifer loafing barn, reroofed in 1990. In the photo is a nice appearing silo, reaching above the barns.

#52

Maryland dairy barn in a beautiful setting; nice farm.

Silos have been a place for chopped corn stalks (green fodder) for cattle. There are many styles built with various materials. In the Western states, long-pit and bunk silos are increasingly built and utilized. Why? Because you can use heavy farm tractors to compress the silage to make it airtight for better curing and keeping quality. But the upright silos are prominently used everywhere. When I was a young fellow, I hired out to neighbors to crawl into the silo (upright) at corn chopping time to scatter the choppings and tromp, tromp, tromp, etc. It helped to pack the ensilage tighter

#53

California hog barn and pens.

Photo courtesy of Fern Eckroat

at that time and the weight of the corn in the upright silo packed it even more as the curing process took place.

Hog barns are found in most states, but they are most prevalent in the greater corn producing states in the Midwest.

#54

Iowa homestead with hog operation facilities.

Corn and hogs really kept that farmer busy. I really think the smell got to him – he is now into soy beans. He is my niece's husband – we occasionally exchange hog pictures and puns.

I've heard claims that hogs are as clean or cleaner than people. It's hard for me to believe, but I sure do hope so. My wife's doctor sewed a "pig" valve inside her heart (the mitra valve). After nine years, it's still there, working away normally. At least, I haven't heard a squeal out of her.

Photo courtesy of Howard Shafer

Kansas gambrel barn with a lean-to attached. It has an oak hardwood floor in the haymow and a bandstand platform. Many a barn dance was held here. Built in 1920.

#55

Coming up is a series of barns converted to various purposes.

#56

Virginia dairy farm converted to a community country church. What better use?

While traveling west of Richmond, VA, we found this church. It struck our best interest, so we stopped, walked inside and found a man practicing on the drums. What had been a barn this church was so immaculate. Take a look!

#57

A Virginia church with a white cross. Was a barn with a manure smell.

Though I never researched it out, I understand that Virginia has around a dozen such barns converted into churches. Many barns are converted to homes.

#58

Yakima, Washington. Colonial home was once a barn.

#59

Thorp, Washington. Second home away from home was once a barn.

#59A

Photo courtesy of Ellen Coffman

Coffman barn near Allen, Kansas. Originally developed by Wiser of Montreal, Canada, for a distillery, using limestone water for good taste. Prohibition of 1918 caused change to a beef barn.

#60

The Barn Motor Inn, Restaurant/Lounge and R.V. Park was once a barn. Prosser, Washington.

#61

Washington barn built about 1974. Place for precinct voting for 18 years.

#62

"FARMER MEETS FARMER"
Washington Barn Precinct #4502.

#63

Central Kansas R.V. Park with restaurant and lounge was once a double peak barn.

#64

Northern Ohio attractive office was once a barn.

#65

New Hampshire antiques novelty shop and maple sugar candies was once an old, old barn.

#66

I was told this Pennsylvania store-station was once a barn. Very nice place for souvenir hunters.

#67

Pennsylvania craft shop. Nice barn appearance and fine handcrafts.

#68

Pennsylvania Amish Transportation – For You – "All Aboard"

#69

Michigan barn renovated for show and a former farm wife's memories. Photo courtesy of Bob Nichols

#70

A Texas Barn/House. A sketch of the "real thing" (house built on top; 24'x24'). Owner furnished a clipping and this writer did the sketching. Photo courtesy of Mervel Lyle

#71

Eastern Washington dairy barn moved and restored with "a great valley view" for a restaurant/lounge or whatever. Built in 1936. Size: 60'x40'x35'

Acreage with barns became public parks:

#72

Kelsey Creek Park, Bellevue, Washington. A park for all ages of people to enjoy any day, all day. A park with domestic animals, arts and crafts, a log house, and three barns.

#73

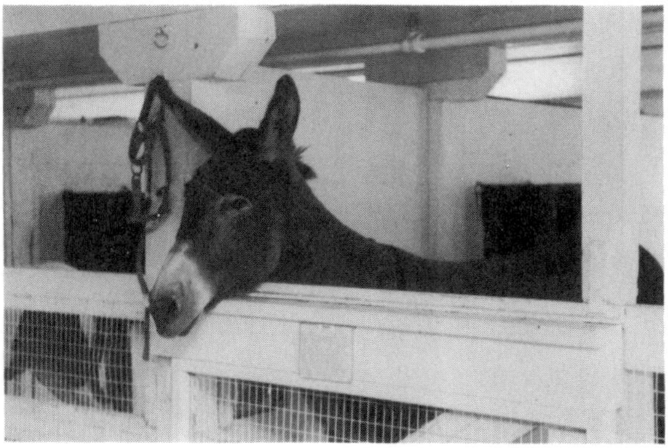

"Pasada" the donkey—the children's favorite animal at Kelsey Creek Park, Bellevue , Washington.

#74

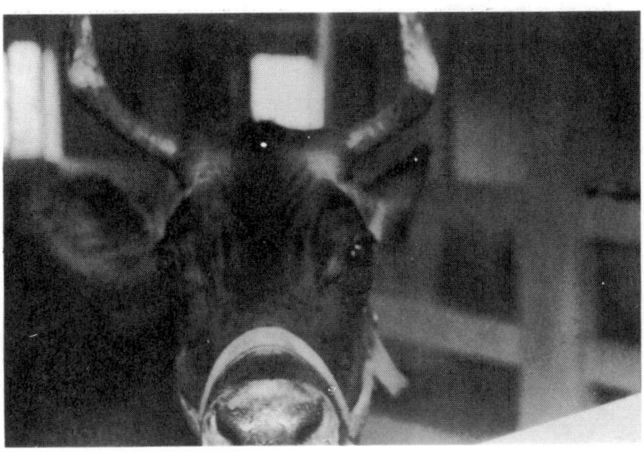

"Beauty" the Jersey cow, the Bellevue street walker. "She is a beauty with age and long horns."

The City of Bellevue had a great vision for bonding family life, for families to enjoy a "day on the farm" with real live, domestic and tame animals; to enjoy a day learning about these animals and their care; to enjoy a day learning arts, crafts and cooking; to enjoy a day picnicking by a creek of clear running water, and to enjoy a day walking the pathways for exercising the body, mind and spirit. It works! It truly is an exciting place to enjoy all of these activities. I love it!

#75

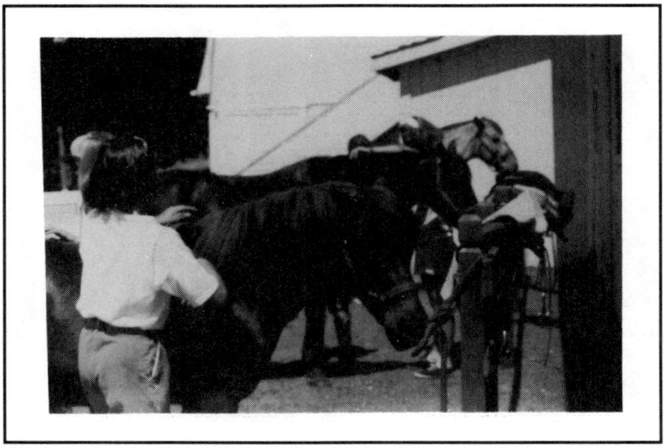

Kelsey Creek (Farm) Park with pastures, meadows and woods. A most picturesque and inviting environment setting for family recreation where an estimated 500 people a day visit during its peak season.

To give you a little background history, I will be brief. The farm goes back to 1921, when it was begun by clearing some stumps and some farm building construction by a couple Mr. and Mrs. W. E. Duey on rented 190 acres of land, where they labored to deliver eastside residents with milk, cream and home-churned butter.

It was not until 1968 when the City of Bellevue then purchased 80 acres, adding another 40 acres in 1990. This made a sizeable area of land to create a park for their plans.

The farm park began in 1972. It was a good thing. The City of Bellevue about that time had a wandering, curious and unidentified "street walker." She was an endangerment to the city traffic, also causing people to "stop and stare." Well, this street walker was an attractive Jersey cow, later given the name "Beauty." Being unclaimed and having no home, the City of Bellevue had the right place for Beauty, the cow. In time, other farm animals, including Pasada, the friendly donkey, were added. Then these animals (including poultry) soon added their off-spring, thus creating excitement and production. Production from many of these animals paid for their keep. On this farm, you could experience what all farmers' children down through the ages have seen many, many times - "birth - life- death."

As for the three barns (and I bet you thought I had forgotten them - no way) they add a scenic beauty and an important service to all who come there. That little barn with a little cupola in the front (on the left), formerly a milk house, serves as a tack shop and animal work office. The middle and large arched roof barn houses the animals. The large arched roof barn (on the left) lends itself nicely for arts, crafts, and cooking. Behind this barn is a playground with children's play equipment - "a fun farm."

For the Kelsey Creek Farm history, activities and stories from which I gleaned this brief, I thank Ms. Tamar White, Farm/Public Relations Technician. She seemed so enthusiastic about the farm and the activities offered, I was greatly impressed. Frankly, it may be a project example for other communities to pursue.

Years ago in New York City, I listened to the "Ink Spots" harmonizing various favorite tunes. One was my favorite "If I Didn't Care." It made a profound impression on me in the days of my youth. In my years of barn research across America, I have learned to know ever so many caring people, really caring about our adequate daily needs; about our love and respect for each other; about giving a helping hand to someone in distress or burdened with a heavy load; about a decent and safe environment; about saving our historical culture; and, yes, let's include saving our barns of the pioneers. These barns are landmarks. It is so gratifying to find community groups organized to save a barn and put it to good use, as the City of Bellevue accomplished. There are other groups around the nation involved in such projects. It reminds us of "a new day dawning."

Keep in mind (positive attitude) that saving the barn is not "the impossible dream."

More group action - saving the barn.

#76

Five Lane Barn at the Hovander Park near Ferndale, WA. Barn built by H. O. Hovander in 1911 (88'L x 66'W x 60'H) Five passage ways through the whole barn - unusual. Now a lovely park where people come to enjoy a day.

A trained architect in Sweden, Hakon Olsson Hovander, migrated to the United States near the end of the nineteenth century, developed a 60 acre farm on the Nooksack River, Whatcom County, Washington. He, his son Hugo, and a crew built a lovely, artistic gingerbread home with 52,000 board feet of native clear fir, hand selected by Hovander. It stands as a beautiful family legacy today. The barn he built is most serviceable with five full-length passage ways with thick heavy floor timbers. Very different, yet attractive. The farm in 1971 was dedicated as the Hovander Homestead Park and placed in 1974 on the National Register of Historical Places, as per Mark Prather, Park Manager.

#77

Barn at Old Town Museum in Burlington, Colorado. Was originally raised near Kanorado, Kansas, beginning 1937 and finished in 1939. (80'L x 38'W x 44'H)

Thanks to Darlene B. Scott, Executive Director, who provided me with this barn information. She told me that the barn was located three miles east and thirteen miles north of Kanorado, Kansas, and it was originally built as a dairy barn with a hay sling in the hayloft (mow), able to hold 180 tons of hay. When the barn was no longer needed on the farm, it was prepared for moving in 1984, taking fifteen days to get it ready with steel beam framing and two days to move it on airplane tires by a Dodge truck to the Old Town Museum, visible from I-70 freeway (Denver to Kansas City). The moving project cost $2,328. Seems most reasonable.

This late model (1937) and well preserved barn has a smooth polished floor used for dances and other community activities. The western village setting has a variety of shops open for business. The barn has a roof style known as the "gothic." Two metal vents on top are there for air circulation. Ms. Scott says, "The city of Burlington owns and operates Old Town." She further says, "Old Town is open year around for tours."

If you leave here on I-70, travelling further on east into Colby, Kansas, you will notice another huge barn. Beginning in 1937, Ben Foster, a lumberman/farmer, built this barn for show cattle. And now it has been moved into town. It towers above all other buildings in the Prairie Museum of Art and History in Colby, Kansas.

I was notified by the museum staff researcher, Inez Mountford, who knew of my barn project, that time and weather conditions were right for moving this barn in May of 1992. She thought this was an exciting and unusual event, one that I might wish to observe and photograph: The moving of the barn 16.3 miles from a farm northeast of Colby to the Prairie Museum Park. Of course, I was interested.

My plans and schedule had previously been made to fly to Denver, rent a car there and research "soddies" (sod houses, barns, schools and churches) in the central states of Colorado, Kansas, Oklahoma and Nebraska. But my plans were for a later date. I quickly changed travel plans, arrived in the Colby Area and was able to spend three days, enjoying this exciting moving event, as hundreds of other people gathered there to see this event.

Here are several photos among many I took, which will give you some idea of the magnitude of this moving task by the movers, Williams Midwest Company of Hastings, Nebraska. Remember, this 150 ton barn is 114 feet long, 66 feet wide and 48 feet high plus two flags on top - the national and state flags. Lloyd White, one of the original organizers with Max

George, stood on a little platform up in the barn peak, raising the flags up and down to be able to get the barn under some of the electrical wires. Of the 21 such wires, some had to be cut. On the last three mile straight away, holes were dug beside wire poles to allow the poles to be shoved over at about a 45' angle for the barn to pass by. All mailboxes and road signs for the 16 miles were pulled up for the barn moving, then replaced by a crew.

I have tried to prepare you sufficiently to enjoy and appreciate these photos. This is the Cooper barn.

#78

Cooper Barn...On The Move

#79 The Cooper barn cutting a swath across the corner of a Kansas wheat field. One other wheat field was crossed earlier for maybe a quarter of a mile with the wheat around knee high. The damage was minimal.

#80

Cut lines allow Cooper barn to move. Hundreds of curious site-seers would come and go, hardly believing what they were seeing. But it was real.

#81 Movin' along. Third day on the road to Colby, Kansas. The road is normal width and makes the barn appear to have a big "hang over." It is doing fine, all in one piece.

#82

The Cooper barn has arrived to its new resting place. Mr. Williams is sighting the barn on to its concrete resting place. Two four-wheel drive tractors with cables helped the 18 wheel truck for this final move.

#83 The Cooper barn is now seated, getting ready for service as a show place for farm tools and other pioneer items.

This is the end of this moving episode, except all wires need to be spliced and tightened, all poles reset, all mail boxes and road signs replaced and all road ditches opened, where they had been filled in. A project worth seeing.

The next huge barn coming up located northeast of Woodston, Kansas, was in earlier years known as the Thomas Barn. The farmer/banker, William Grant Thomas, hired a local architect/builder, Robert I. Brittain to erect the barn, completing it in 1912, according to Lee Hull, farmer and past president of the new restoration project. It measured 100' long x 64' wide and 64' high with the capacity for 500 tons of loose hay. The barn was designed and built to hold 54 horses and 50 head of cattle, costing $8,000 to build at that time. It had a gambrel roof with 2 large cupolas on top and 4 dormers (two dormers on each side of the roof) for light and air circulation. This barn was built with 120,000 feet of lumber, 115,000 shingles, 7,000 pounds of nails and 1,200 pounds of bolts. What an interesting record, isn't it?

Farm tractors and large implements gradually replaced the horses and small horse drawn equipment, leaving the barn to become an idle structure on the ridge. Talk about change, this was a revolutionary reality all across American farmsteads, following World War I and rapidly increasing through the 1920's and 1930's.

A visiting minister to the community for a 1988 annual memorial service in the stone church, Reverend Richard E. Taylor, Jr., of Topeka, Kansas, marvelled at the mammoth size of this unusual barn, had a second take (so to speak), and envisioned some future use of it again. A non-profit corporation, known as the Classic Big Barn, was formed to raise funds and restore this so called "Pride of the Prairies." This Classic Big Barn was dedicated in 1989 and the race was on, raising funds across Kansas and other states with an effective preservation appeal to keep the barn from "spreading its sides and biting the dirt." The word went forth into the land.

This author, being in earlier days a Kansas "Jayhawker" boy, with dirt gone now from under his fingernails, but with long lasting memories of those farm days, not far east of the location of the Classic Big Barn, was tempted to join the group as a charter member, along with his wife, and pay the dues of restoration of a barn. Something he had urged others to do in other places, as he and his wife traveled across America with a passionate campaign; "Let us Save the Barn." Well, here is the photo of that very barn. It is a pride to me.

#84

The Classic Big Barn, Woodston, Kansas.

This barn and 15 acres were donated by Clifford Thomas, son of the builder – thanks to him. The barn is listed on the National Register of Historical places. When the haymow is restored, hopefully this year of 1993, the hope and possibilities are for the barn becoming the Barn Dance Capitol of Kansas. Donations are needed and welcomed. Will you help?

Kansas, being heavily laden in many counties with limestone rock, lays claim to hundreds of limestone buildings: homes, churches, schools, courthouses and barns. Here comes one.

#85 The Auld Stone Barn, south of Wakefield, KS. Photo courtesy of Jeanette Braden

This limestone barn was begun in 1908 and finished by James Auld in 1910. It measures 180' long and 30' wide. The location is 5.5 miles south of Wakefield, Kansas. Jeanette Braden tells me that hired masons cut most of the regular shaped blocks but, not knowing the technique of cutting arched blocks, her father cut them. Stone barns can take a lot of strong winds.

#86 The Berthusen Park Barn, northwest of Lynden, WA.

Here is a northwest Pacific coast barn, located in a very popular community park, called the Berthusen Park, which is about three miles or so northwest of Lynden, Washington. The original smaller part was built in 1887, measuring around 40' long and 24' wide. In 1901, Hans Berthusen built onto this small barn a much larger barn, measuring 188' long x 128' wide x 50' high. It was built of split timber from giant cedars in that area. Notice the arched door ways in the right side of the barn. That is the Dutch influence, as that area was settled by many Dutch farmers.

The manager of Berthusen Park, Earl Nydam, has generously provided me with information on the barn history and current use of the park. Hundreds of people come to the park during the first week of August for the Annual Puget Sound Antique Tractor and Machinery Show. It is a great show of old tractors (even the steamers), farm machinery and the antique one-lunger gas engines.

I love to go there each year. The show, the parade of old trucks, tractors, oddities, plus the food - it's hard to resist.

In the eastern United States I found many of the older standing barns had mortise and tenon joints pinned together with wooden pegs. Likewise, the older western coast barns are built the same way, especially those built in the 1800's. This method does make for a strong structure. Here is one.

#87

Photo courtesy of Dale Rutledge

This Rutledge barn southwest of Tumwater, Washington, was built by Thomas Rutledge in 1864 with hand hewed timber, mortice and tenor joints and pinned with wooden pegs. (124'L x 74' W x 36'H)

Remember the movie actor Walter Brennan, now deceased, who was very popular in western movies? He loved the northeastern Oregon mountains and Wallowa Valley, where he owned a ranch with a gambrel roofed barn with his branding iron symbol painted on the north end. As you look at this photo, you will probably not guess its meaning. It represents the O. K. Quarter Circle Ranch. It is obvious to you now, isn't it? Walter's son, Mike and family live on the ranch, southeast of Joseph, OR.

#88

The Brennan Ranch near Joseph, Oregon, in the lovely Wallowa valley.

No doubt you have heard of houses connected with the barn, with a passage way through the summer kitchen, through the workshop/tack shop, through the buggy/wagon barn, and then to the animal barn. In varying combinations, these arrangements are found in Europe and in the New England states in the United States.

This is a Swiss connected barn and house of an older style. The next one is more modern.

#89

Photo Courtesy of John & Irene Peters

#90

Another Swiss connected barn and home in Switzerland in a more modern style. It has a dirt ramp up to the haymow at the barn end.

#91

New Hampshire connected barn and house with long passage way inside.

These types of structures are fine for winter when tending the livestock - no walking in the deep snow.

They tell me that there is a problem in summer with an abundance of flies. Advantages – What is best?

Occasionally you see a courtyard complex of barns. This one happens to be Canadian.

#92

Canadian courtyard complex of barns.

This courtyard complex of barns had one gate opening to allow the animals passage to and from the pasture. I have heard of completely enclosed courtyards, but I have never found any. Even this style or arrangement is not too common here in the United States. There are advantages - the livestock is protected from the strong, cold storms.

Pennsylvania Dutch barns are usually decorated with various circular symbols on their barns. I will identify a few here: Oak leaves represent or symbolize "strength - character"; the Irish hex with three leaf clover and two birds represents "good fortune"; Willkommen with two birds looking in opposite directions to see who is approaching represents "hospitality"; the rosette represents "love-romance"; and the unicorn represents "peace-contentment". Here is a California barn decorated.

Photo Courtesy of Fern Eckroat #93

The builder had been to Pennsylvania, saw the Pennsylvania Dutch barns with the circle decorations, liked them, returned to California and added them to his big barn.

#94

The one-horse barn near Mt. Vernon, Oregon, built of stone with two gun portholes in each of the four sides. Designed to protect a special horse named "Mt. Vernon" from potential thieves.

Round Barns...

Round barns are sometimes referred to as circular barns or polygonal barns. A polygonal barn usually has over four sides or angles, yet appears to be round compared to a rectangle, square or L-shaped barn. My research on round barns is extensive enough to warrant a separate book for them in another publication, expected to be published in 1994 or early 1995.

#95

The Sue Marble round barn located south of Grandview, Washington. This photo includes the author of this barn book, telling a friend about the different styles of round barns.

The parade of barns with photos is herewith completed. However, there is more important commentary to come.

Please read on...

Chapter Five

Barn Therapy

Barns need therapy from time to time, as we do, to maintain a healthy, productive life. This chapter attempts to help a barn owner to do just that with anyone of several problems.

Before his retirement on December 31, 1991, from the Department of Agricultural Engineering at the University of Idaho, my friend, Roy Taylor, had been providing instructions to farmers and ranchers to help them build or rehabilitate their buildings. Upon request, Dr. James A. DeShazer, head of that department, gave me his written permission to use some instruction material for this book on barns, to further help other farmers and ranchers with their buildings. Roy Taylor wrote *Balanced Constructions in Farm Buildings* and I shall present it verbatim.

"What do many farm buildings and a full-back with a gimpy knee have in common? — a weak point, a flaw, that limits the capability of the total mechanism. Six feet four inches of highly conditioned body and 240 pounds of meanness cannot compensate for a torn ligament.

Balanced construction means that all structural components have approximately the same capability to withstand those load stresses that the building is designed to carry. A 40 pounds-per-square-foot (spf) load carrying capacity in the roof structure does little good if a building load of 20 psf is going to break the footings.

Imbalance in building construction means that extra dollars and extra time have been spent fruitlessly. The most serious consequence however, is that the load carrying capability of the building is misleading. This in turn can, and sometimes does, lead to structural failures requiring expensive repairs of replacement of the building; that damage or destroy equipment, machinery, crops, or livestock that are housed in the building and that pose the threat of injury or death to human occupants.

A typical farm building is a complex series of component parts, fastened together for the purpose of providing some degree of controlled inside environment. Each component part, whether it be a foundation wall, a pole, a beam, a stud, or a rafter must be capable of carrying its share of the total building load. Likewise, the connecting joints that hold these component parts together must be capable of carrying the loads transmitted through them.

A correlation exists between the structural members of a building and a human skeleton; the joints are the most susceptible to damage or failure. Unfortunately, it is these joints that are often most neglected in on-farm construction.

Loadings on farm buildings take many different forms. the most obvious are the loads pressing vertically downward: the weight of the structure itself; the weight of accumulated snow on the roof; and the weight of animals, equipment, and produce that are in or on the building. Horizontal loadings result from wind and from unequal fluid loads (stored grain for instance) that push outward against a wall. These sideways forces tend to tip buildings over or to buckle the roof-to-wall and wall-to-foundation connections where the greatest leverage occurs. Wind also creates a vertically upward force which sometimes pulls buildings up by the foundations or tears off roofs.

Open faced buildings, or any building in which large doors have been left open, are subject to internal wind pressures which push outward against walls and upward against roofs.

Impact stresses occur from moving or bumping loads: a truck or other pieces of equipment bumping against a door frame or large animals crowding against a wall. Snow sliding off the roof causes a combination of vibrating horizontal and vertical stresses, particularly if there is a drop-off from one roof level down to another as often exists in livestock/hay storage buildings with shed roofs protruding from the sides.

In evaluating a building for balanced construction, each structural member and connecting joint must show the capability to withstand stresses acting downward, upward, and horizontally.

The most important thing that can be done to achieve a reasonable balance of construction is to strengthen the weak spots. Beef-up those structural components or joints which are responsible for most building failures.

Farmers and ranchers should know the snow-load and wind-load potential for their locations. county building code officials or Cooperative Extension Services offices can help provide this information.

Truss or rafter systems should be well engineered. Unless special care is taken in on-farm construction it is best to purchase trusses for wider spans from a dependable manufacturer.

All structural lumber should be carefully selected to make sure it is sound and of a grade A species, Douglas Fir or Larch (Tamarack). Other structural species have about 40% less strength and require special designs.

Hardened steel screw-shank or ring-shank nails produce holding power nearly equivalent to wood screws. Even with these nails it becomes increasingly important with wider spans to use glued or bolted connections on trusses. Nailing patterns to support heavy loads are very critical to achieve adequate gusset/timber connections.

Most foundation problems under wood-frame or concrete block walls can be prevented by using minimum amounts of steel reinforcing bar. The rebar helps eliminate shrinkage cracks and corner separations and provides stiffening for bridging over soft spots in the earth beneath the foundation. Special engineering should be applied to foundations which are placed in unstable, poorly drained soil.

Floating foundations those which do not go down below the frost line, work well for buildings such as metal grain bins if the entire concrete slab floor is reinforced together so it "floats" as an integral unit.

Pole or post construction results in concentrated ground loadings due to their wide spacing. Footings beneath piers should be equal in total contact area to that of the normal wall foundation it replaces. Preservative treated poles or posts require less base footing area due to the gripping resistance of compacted backfill around the buried portion. Poles or posts which are set too shallow do not have enough surface contact area against the ground and are apt to tip or sink under sustained heavy loading. They are also more apt to pull out due to wind uplift forces.

Maximum resistance to the failure of horizontal-to-vertical connections can be achieved only by providing triangulation to form rigid joints. Triangular bracing with knee braces or gusset plates also increases the load carrying capacity of the horizontal and vertical members to which it is attached.

Poles or posts which are bolted to piers or to a concrete slab form "hinged" joints. The limiting factor with hinged joints is the tendency of the connection to break due to the extreme leverage working against it. The top of the pole, if fastened only to the roof members with which it comes in contact, is also a hinged joint and has the same limitation.

A typical pope or post which has one rigid end has about 2 times the carrying capability as one with hinged joints on both ends. A pole or post which has both ends rigid can carry about 4 times the load of one with hinged joints on both ends.

The plates which connect between the tops of poles or posts are one of the most neglected components in that type of construction. They often must carry the heavy concentrated loads of rafters or trusses along their span. These members should not only be carefully matched against the loadings they must carry, but also need special support to hold them to the pole top. They should either sit on top of the pole, be notched into it, or be held by a support block fastened to the pole beneath them. Bolting or strapping of the plates to the pole is essential. A common cause of pole building failure is the stripping of these plates from the poles or posts which were to hold them.

Another often neglected connection is that of rafters or trusses to the top of the wall. Downward, upward, and lateral forces become concentrated at these points. The best and easiest method of anchoring is by using plumbers strap or special metal clips. However, wooden anchor blocks securely fastened to both the wall member and the rafter/truss member can suffice.

Concrete block walls very often crack. They are heavier than wood-frame walls and therefore place more stress on their foundations. They also do not have the flexing capability of wood, and even a slight degree of settling or other building movement results in fracturing of the masonry. Footings and foundations beneath concrete block walls should be reinforced, and the wall itself should be tied together with steel rods and wire mesh.

This article has been directed mostly towards wooden buildings since wood is the material most often used in on-farm construction. Wood is perhaps the most forgiving of all construction materials. It has the capability to withstand considerable overloads, particularly if they occur for only short periods of time. It it were not for this placable characteristic of wood we would see a great many more farm buildings damaged or destroyed by extremes of snow and wind.

By recognizing the benefits of obtaining balanced construction and by learning to recognize the weak links in the structural framework of a building, the satisfaction and the safety of a farmers investment can often be substantially increased.

Buildings are complex combinations of structural components. the larger the building, the more complex and the more important each component becomes. Each type and size of building at each specific building site requires somewhat different considerations. To obtain information on general construction methods and to obtain plans for specific farm buildings contact your county Cooperative Extension Service office.

In addition to Taylor's instructions, he provided through the Idaho Farm Building Plan Service helpful information from the Republic Steel Corporation, *How to Repair Damaged Farm Buildings*, for which we express thanks and give credit of acknowledgment. It lists causes of failure and steps of repairing in reference to the foundation, sill, siding, sagging roof or a tilted building.

HOW TO REPAIR DAMAGED FARM BUILDINGS

Undermined Foundations are most often caused by water draining off roofs, washing away and weakening the soil supporting the building. This rots the framework and sometimes causes moisture difficulties inside the structure. A permanent remedy includes running a concrete strip completely around the building, sloping it downward to carry the water quickly away from the foundation.

Undermined Foundations

Causes of Failure:

1. Erosion or a hog wallow caused by lack of drainage;
2. Lank of support under footing at a corner;
3. Lack of reinforcement in the top of the wall.

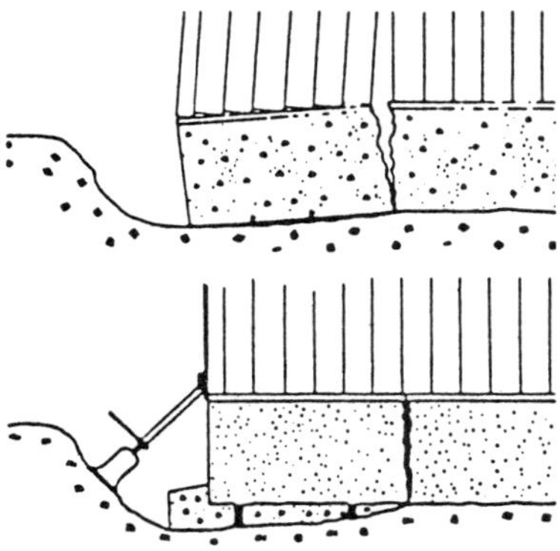

Steps in Repairing

1. Dig under foundation at one point to provide room for a jack;
2. Chip off about 2 inches of the wall at the break to aid in getting it back in place;
3. Lift foundation and building back into place;
4. Wedge bricks or cement blocks in under the wall, as close to break as possible;
5. Build a new concrete footing along the outer wall, tamping it well under the old wall.

Foundation Split at Corner

Causes of Failure

1. Pressure on inside of both walls causes extra load on foundation at corner;
2. Lack of reinforcement at corners; 3. Excess water outside foundation at corner;

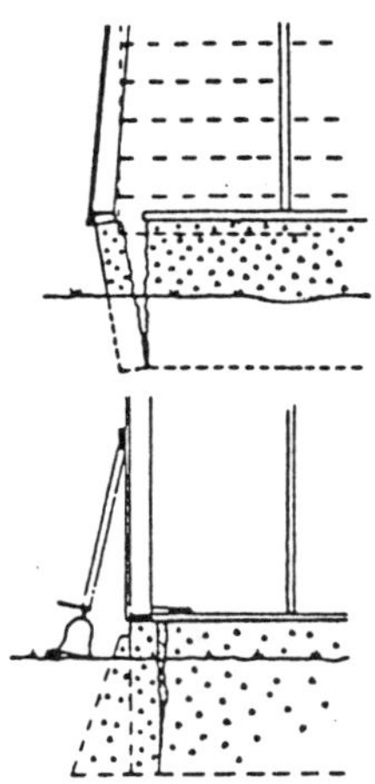

Steps in Repairing

1. Chip away about 2 inches of wall at break to allow end foundations to come back into place;
2. Jack wall and foundation back into correct place;
3. Build concrete brace along old wall to hold it in place;
4. Nail a "Tie" board across corner on inside to hold sills in place;
5. Patch opening in foundation wall. When building a new concrete foundation wall, use reinforcing steel rods at corners to avoid this kind of break. The rods are bent at right angles so that they run right around the corners. They are placed in the concrete while it's being poured, about 6 inches from the top of the foundation wall.

Foundation Wall Titled Out

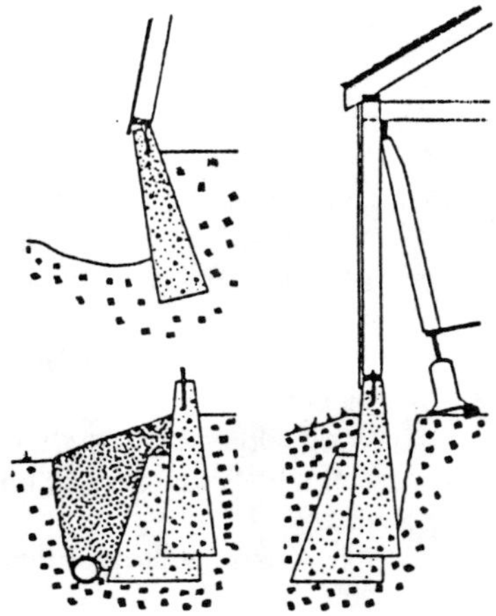

This is a common type of foundation trouble. Good roof drainage, carrying water away from the foundation, would probably have prevented it, for wet soil has about one-half the strength of dry soil.

Causes of Failure:

1. Soil washed away from beside and under foundation, or soil loosened;
2. Pressure was developed inside of foundation wall;
3. Livestock may have crowded or pushed against wall;
4. Footing was too small.

Steps in Repairing:

1. Jack up one section of the building at a time; raise slightly higher than normal position;
2. Straighten foundation wall and pour new footing as shown;
3. Allow time for curing - then lower section into place;
4. Provide drain spout - or slope the soil around building so that water quickly runs away from building;
5. A tile drain, covered with gravel, may be used instead of spouting.

Siding and Sill Rotted

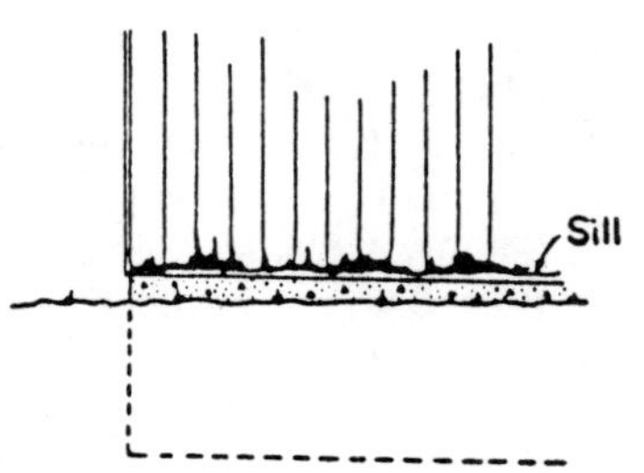

To avoid recurrence of this trouble, the most practical solution is usually to build up the foundation wall so that the building framework is raised far enough above ground to escape moisture.

Causes of Failure:

1. Sills and siding too close to ground to dry out readily;
2. Rainwater splashed mud on siding.

Steps in Repairing:

1. Lift a section of the building higher than its final position;

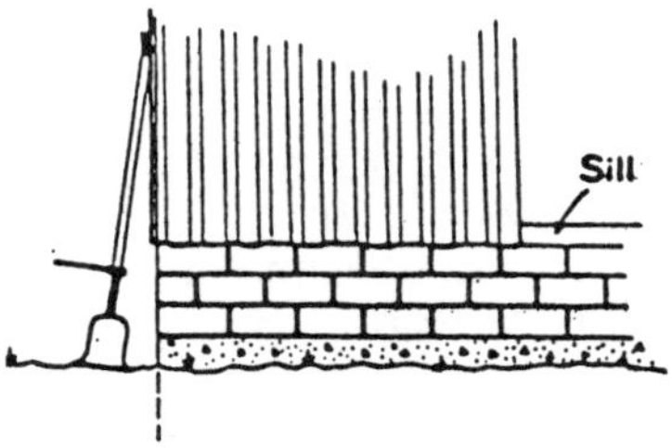

2. Saw through siding and studding at a height that will permit the desired number of rows of blocks, tile or bricks will fill in;
3. Lay up new wall on old foundation and allow mortar to set;
4. Install new sill board on top of foundation wall and lower building into place. Make sure that studs and siding are siding with pressure-treated lumber or Rigid-Rib galvanized steel siding.

Siding Repairs
Repairs to damaged sidings should be made as quickly as possible after discovery to prevent damage to floor and sill as well as spreading of the damage to other siding areas.

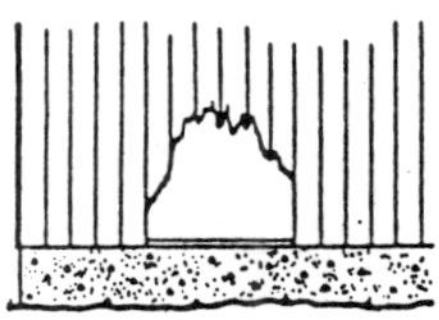

Causes of Damage:
1. Siding boards broken by livestock;
2. Lack of protection on inside of building;
3. Insufficient room back of stalls.

Steps in Repairing:
1. At a point above damaged section, square across broken boards and saw along this line to even up broken ends;
2. Fasten a metal strip along the ends of the old boards to weatherproof the joint, bending the strip over the ends and fastening again from the inside;
3. Nail a board across the top of the opining from the inside, letting it extend down far enough so that the new siding can be nailed to it;
4. Saw new siding boards to length and nail in place.

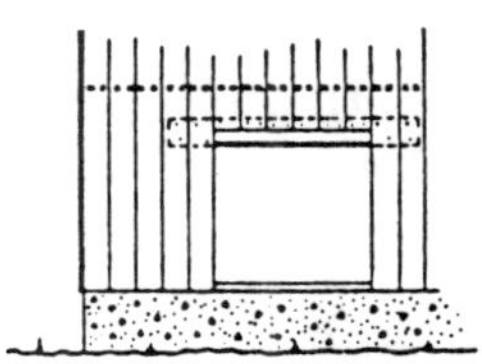

Sagging Roof

This type of trouble can rapidly grow worse if remedial measures aren't promptly taken. The likelihood of a sagging roof can be greatly reduced if the frame is tied in to prevent spreading as shown in the bottom or middle drawing.

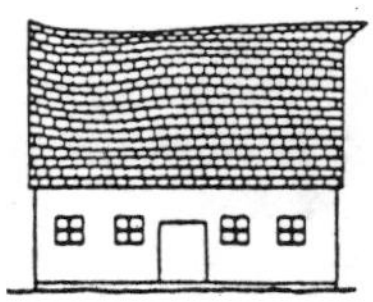

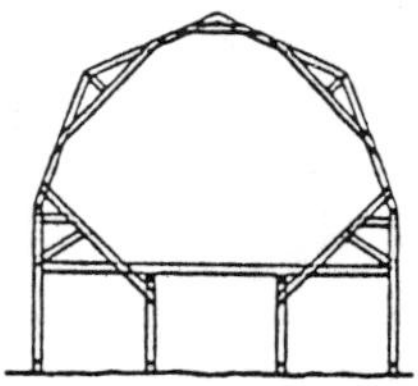

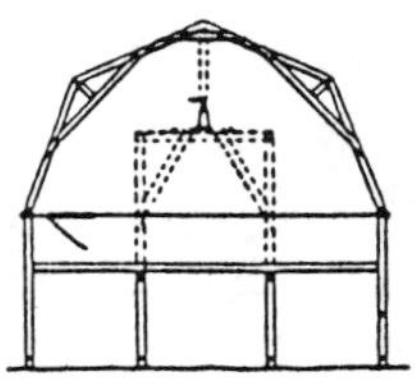

Causes of Failure:

1. Pressure of hay on frame and on sidewalls;
2. Lack of stiffness in plate;
3. Incorrect bracing.

Steps in Repairing:

1. Build up a platform from which to lift at center;
2. Attach a cable and block or use a rod and turnbuckle to pull sidewalls in at the plate level;
3. Apply both at the same time to straighten the roof line;
4. Install side braces made of two 2" x 8" pieces at intervals of 8, 10, or 12 feet;
5. Renail or bolt all braces affected.

Tilted Buildings

Small structures with large doors at one end, such as garages, will sometimes lean to one side. Adequate bracing is the remedy - and the sooner it's done, the better.

Causes of Failure:

1. Doors are open a great deal and wind causes strain;
2. Lack of wall width to give stability;
3. Weight of doors causes strain.

Steps in Repairing:

1. Straighten building until sides are plumb; 2. Nail short tie to hold rafter in place; 3. Nail brace to rafter, joist and studding. Set brace as far down as possible without interfering with door movement; 4. Nail and brace across rear end if needed.

Author's Comments: If solutions to your questions or problems have not been answered thus far, I would suggest an inspection by an engineer or builder to determine the extent of damage, deterioration, or weakened joints. It is possible this inspector could provide optional solutions for whatever problems may exist. Also, I have checked various libraries, and you may wish to check your local library. Often they have construction books on hand and listed in their card file or computer for your information. Sometimes, it is difficult for libraries to keep current, especially on the rising cost factors. But it may pay you to check.

Chapter Six

A Barn Man's Biography

HAROLD E. WAGNER
author and barnologist

Harold was born April 11, 1917, in the Frank A. Wagner family on a Kansas farm; was taught discipline and hard work; was college educated in business principles, liberal arts and theology; had training in the military service and served 4 1/2 years in the U.S. Navy during World War II. He was awarded a B.A. degree in World History, a Masters degree in Theology, and numerous business seminar certificates; had a career in farming, ministry, and business plus an avocation in carpentry, volunteer charity work, photography, artistic painting, and writing. He married Olive Burnette on June 15, 1947, and together they have three grown children and nine grandchildren. Harold Wagner has held membership and has been active in church, civic, business, agricultural, social, poetical and historical organizations. His philosophy has been "A person's positive attitude and true wealth are the good deeds a person does in this life," as reported by his wife, Olive.

What is a "Barnologist"?

The above biography indicated the extent of training, education, and experience during my longevity of my working life before retirement. In retirement years, rocking never appealed to me, but several areas of an avocation caught my attention and gave me new challenges. Pioneer barn research surfaced. It needed a scientific term for identification.

Various terms were considered, but coining the word "barnologist" seemed most practical, applicable, and identifiable of a serious research of barns. It denotes one who studies barns: the different purposes for the barns, the structural parts/terms, the various styles, the constructional procedures, rehabilitation, and maintenance.

This new word has not reached the Webster Dictionary or any other such book, as per my research and for years I have used the term. Words are born daily, so it seems. Just listen to our youth with their new vocabulary. Their creativeness is amazing.

When my wife and I on October 3, 1990, visited Mary Humstone, Project Director, National Trust for Historic Preservation, in her office in Denver, Colorado, we wanted to learn more about the "Barn Again" program. It was relatively new at that time; creatively rehabilitating barns; bringing renewal, stabilization and longevity to barns - a commendable project.

As I handed Mary Humstone my business card, she curiously inquired where the word barnologist originated and what it involved. She said, "The term was very appropriate."

Several friends across America have expressed an interest in forming a Barnology Club to share concerns and solutions through a news network to the members. What do you think? Would it be helpful?

Let Us Save The Barn

by Harold Wagner

Farmers built many barns in the pioneer days
    To house their livestock, hay, and grain.
Those barns were built in such various ways
    To protect against the winds, snow, and rain.

Why were some of those barns built so round?
    Central feeding was the greatest reason,
And they attracted public attention, I found;
    Also, their practicality served the season.

Most barns are rectangle in their size and shape,
    Lending the square footage for better storage.
It was easy to add lean-tos at some later date
    For accommodating more livestock and forage.

Why were the painted barns first painted red?
    It's a good question that has often been asked.
Iron oxide, skim milk, and lime were all they had,
    Until later, when multiple colors were amassed.

In earlier days, wood or stone were used for barns.
    During World War II metal Quonsets were built,
Which led to metal pole barns for the many farms,
    And stored tractors replaced horses to the hilt.

Now, across America barns are falling with decay.
    Some are changed to an office, church, or home;
Others are being destroyed, causing great dismay.
    Then let us save the barn and end this poem.

**"Faith in Man and the Way,
Hope for Man and the Truth,
Love of Man and the Life."**

H. E. W.

But you may ask, "What does faith, hope and love have to do with barns?" That is a fair and profound question. Faith, hope and love are "motivators," giving meaning and impetus to "live and let live."

Without these three motivators, I would never have considered expending the thousands and thousands of hours, miles, and dollars to produce this book.

ACKNOWLEDGMENTS

In one way or another, those of you whose names are listed here, have had a part in the production of this book. I thank you.

Jeff Andrews
Larry and Carol Ashby
Laura Aymond
Frank Bailey
Ardath and Mark Beringer
Jim and Marilyn Berry
Berthusen Park, WA
Mildred Biles
Gerard Billington
Viola Blaylock
Gerald "Jerry" Boren
Jeanette Braden
John and Myra Branscom
Kim Brewer
Don and Velva Butler
Mark Callender
Carnation Company, WA
Canterbury Shaker Village, NH
City of Bellevue, WA
City of Burns, OR
City of Hays, KS
City of John Day, OR
City of Mt. Vernon, OR
City of Red Cloud, NE
City of Stockton, KS
Girard and Betty Clark
Cecil and Polly Cook
Ellen Coffman
George Colburn
Don Cole
Gary and Elfriede Cooper
Department of Agricultural Engineering - Unv. of Idaho
Dr. James De Shazer, Jr.
Charlotte Dixon
Ruth Dirk
Cathy Druell
Wilbur Duim
Linda Dunham
Gala Eaglin
Fern Eckroat
Kenneth "Kenny" Eliason
Dean Elliot
Ardis Emery
Charles Eshelman
Tom and Toni Fest
Wayne Fields
Elizabeth Fitzimmons
Gordon Flint
Helen Frahm
Paul Fridlund
Henrietta Gavin
Max George
Erma Gfeller
Dr. Joseph W. Glass
Joe Goodman
Lawrence Goodnight
Grant County Chamber of Commerce, OR
Betty and Myron Grubb

Hancock Shaker Village staff, MA
Frank Hansen
Elmer "Bud" Hart
George Hathaway
Jim and Carol Hartman
Harold and Wilma Heitschmidt
Hovander Homestead Park, WA
D. Lee and Kelley Hull
Lawrence and Marvel Hull
Mary Humstone
Mike Hurley
Liz Johnson
Joseph Museum staff, OR
Kelsey Creek (Farm) Park, WA
Pat Kezele
Harvey Knoll
Carol Lange
Maury Lange
Sue Lavelle
Lewis Alumni Center, W.S.U.
Cliff Lewis
Leslie and Bertha Linville
Keith Lincoln
Jan Lowry
Mervel and Roy Lyle
Gabe Lyons
Malm's R.V. Resort, KS
Sue Marble
Norbert and Rose Marquez
Charles E. Mattocks
Peter McDonald
Kenneth "Ken" Mitchell
Inez Mountford
Mulalleys Art Gallery, ID
Julie Murphy
Dale Mustoe
Mt. Lebanon Shaker Village, NY
Robert "Bob" Nichols
Allen Noble
Earl Nydam
Steven Nys
Robert Offerdal
Elois Ogden
Old Town Museum Park staff, CO
Dr. Leo E. Oliva
Andy and Marion Owens
Duke and Linda Pappenheim
John and Irene Peters
Roger Pond
Jon Pope
Mark Prather
Elsie Reich
Republic Steele Corporation
Winifred and Cleo Reutter
Cay Rhoads
Madeline Richardson
Timothy "Tim" Rieman
Kathy Ruggles
Dale Rutledge
Mike Swanson
Ted and Kay Salverda
Pat Schuyler
Darlene B. Scott
Barbara (Parker) Serafini
Howard and Freda Shafer
Arlo Spore
Dean Squibb
Lester Steele
Ralph and Pat Steele
Chris and Denise Svendsen

Rev. Russel Tate
Rev. Richard E. Taylor, Jr.
Roy Taylor
Sue Ellen Taylor
Pat Temple
Thomas County Prairie
Museum, KS
Phil Thomas
Ila Timson
Marvin Tjoelker
JoAnn Wagner
Mark Wagner
Olive Wagner
Orval and Florence Wagner
Kathrina Warren
Lester Warriner
Dr. Robert Wayne
Webster County Historical
Museum staff, NE
Ernest Weiss
William "Bill" Werst
F. J. Wetherell
Lloyd White
Shirley Willard
Gary Williams
Shirley Willoughby
Gary Younkin
Joe Young
Helen Zwink

NOTE: If your name has not been listed here, please forgive me. Please let me know. If you contributed something about round barns, log barns, sod structures, silos, windmills, out-houses, corn cribs, or about a barn that I received information and/or a photo before the cut-off deadline for this current book, my plans require more research on the round barns, etc., therefore they have not been included in this publication. Your credit will be given in the next book and I do thank you for your patience and participation. Hopefully, I can get the next book published in 1994 or at least in 1995. Sooner the better for all of us. Notice will be sent, following the publication.

PHOTO INDEX

BARNS IN WAITING

BARNS ALIVE – ON DUTY

#61. Washington barn built about 1974 by Wilbur Duim, p. 45.

#62. Washington Barn Precinct #4502, p. 45.

#63. Central Kansas barn converted into a R.V. Park with restaurant and lounge, p. 46.

#64. Northern Ohio office was once a barn, p. 46.

#65. New Hampshire barn converted into Serafini's antiques novelty shop, p. 47.

#66. Pennsylvania barn converted into a store/station, p. 47.

#67. Pennsylvania crafts shop, p. 48.

#68. Pennsylvania Amish Transportation, p. 48.

#69. Michigan barn renovated for show and a former farm wife's memories, p. 49.

#70. A Texas Barn/House, p. 49.

#71. Washington dairy barn moved and restored , p. 50.

#72. Kelsey Creek Park, Bellevue, Washington, p. 50.

#73. "Pasada" the donkey, at Kelsey Creek Park, Bellevue, WA, p. 51.

#74. "Beauty" the Jersey cow at Kelsey Creek Park, Bellevue, WA, p. 51.

#75. Kelsey Creek (Farm) Park, Bellevue, WA, p. 52.

#76. Five Lane Barn at the Hovander Park near Ferndale, WA, p. 54.

#77. Barn at Old Town Museum in Burlington, Colorado, p. 55.

#78. The Cooper barn moving. The barn was donated by Gary and Elfriede Cooper to the Prairie Museum of Art and History in Colby, KS, p. 57.

#79. Cooper barn cutting a swath across Kansas wheat field, p. 58.

#80. Cut lines allow Cooper barn to move, p. 58.

#81. Third day on the road to Colby, KS, p. 59.

#82. Cooper barn at its new resting place, Prairie Museum of Arts and History in Colby, KS, p. 59.

TO PURCHASE A COPY OF THIS BOOK

See your local book store

or

Mail this form (or a photo copy) to:

H & O PIONEER BARNS PUBLISHING
508 E. 3rd Street
Grandview, WA 98930
PH: (509) 882-1799

or

NORTHWEST PUBLISHING
1130 Columbia Avenue
Sunnyside, WA 98944
PH: (509) 839-2931
FAX (509) 839-3455

Please send my copy(s) of "Barns on Parade Across America"

[ ] Soft cover @	$ 19.95 ea	=	$______
[ ] Hard cover @	$ 26.95 ea	=	$______
		Sub Total	$______
		WA residents add 7.6% sales tax	$______
		Add $2.90 each book for Shipping & Handling	$______
		TOTAL	$______

NAME

STREET ADDRESS

CITY/STATE/ZIP

AREA CODE/PHONE

(Please enclose check or money order. Allow 4-6 weeks for delivery.)